SCIENC MUSEUM
L DON

The Wri ht Brothers

A brief c unt of their work

9—1911

By

CHARLE ARVARD GIBBS-SMITH

M.A. (vard); Honorary Companion
of Royal Aeronautical Society

"It is therefore incontestably the Wright brothers alone who r ved, in its entirety, the problem of human mechanical flight . . . Men of genius—erudite, exact in their reasoning, hard workers, outstanding experimenters, and unselfish . . . They changed the face of the globe."

CHARLES DOLLFUS

LONDON
HER MAJESTY'S STATIONER)FFICE
1963

THE WRIGHT BROTHERS
Wilbur (1867—1912) and Orville (1871—1948)

On December 17th, 1903, the Wright brothers were the first men in history to make powered, sustained and controlled flights in an aeroplane; both the machine and engine were of their own design and construction. The longest of four flights lasted 59 seconds and covered over half a mile through the air. This achievement followed three years during which the brothers had developed and perfected the design, construction and pilotage of gliders.

In 1904, in the second of their powered "Flyers" (as they called them) they made about eighty brief flights—including some turns and a few complete circles—the longest lasting just over five minutes, in order to discover and master the basic techniques of controlling a powered machine.

In 1905, the Wrights completed and flew their third Flyer, which was the world's first practical powered aeroplane; during more than forty flights, the machine was repeatedly banked, turned, circled, and flown in figures of eight; on two occasions it exceeded half an hour in flight duration.

It was in 1902 and 1903 that illustrated information about the Wrights' gliding of 1900–02 reached France, and directly precipitated the first stage of practical European flying, at a time when European aviation was virtually at a standstill, following Lilienthal's death in 1896.

In August of 1908, Wilbur Wright started flying in France, and thereby directly precipitated the second—and decisive—phase of practical European aviation; this resulted chiefly from his masterly demonstration of flight-control, especially control in roll, and lateral control in general, which the slowly developing Europeans had neglected.

The Wright brothers were the sons of a United Brethren Church bishop, and lived at Dayton, Ohio. They had progressed from the selling of bicycles to their manufacture, and thereby made a comfortable living; this business alone provided the funds for their aviation. It was the work of the great German gliding pioneer Otto Lilienthal, and his dramatic death in 1896, that first focused their mature attention on aviation. In their subsequent study of bird flight, and of Lilienthal's gliding, they became fully alive to the inefficiency of the latter's method of seeking balance and control solely by body-movements, as well as to the bird's effortless mastery of these problems. Then, in 1899 the Wrights made their first decisive discovery and their first decisive invention: they observed that gliding and soaring birds—evidenced especially by their local "expert" the buzzard—"regain their lateral balance . . . by a torsion of the tips of the wings"; and they decided to apply this bird-practice to aeroplane wings. At first they thought of variable incidence wings, but abandoned the idea for structural reasons: then they hit on the technique of a helicoidal twisting of the wings ("warping"), after toying with a long cardboard box. The Wrights' "basic idea was the adjustment of the wings to the right and left sides to different angles so as to secure different lifts on the opposite wings" (Orville): the bare idea of such control in roll had been previously envisaged, but had never been applied successfully in practice, and never even been imagined in combined action with the rudder to counteract

1. Wilbur Wright.

2. Orville Wright.

the increased drag on the wing with the increase of angle, as the Wrights were soon both to envisage and accomplish successfully.

By character and temperament the Wrights were admirably equipped for their work, being possessed of modesty, pertinacity and a richly inventive talent: in addition, they trained themselves to become both expert engineers and pilots, concerned equally with the theoretical and practical aspects of aviation. They drew their initial inspiration from Lilienthal, and later absorbed an extensive knowledge of the history and state of their craft; but when it came to the specific problems of design and construction—including propellers—and pilotage of both gliders and powered aeroplanes, they had to unlearn most of what they had learned from others, and start afresh.

Mention should be made here of occasional—but vociferous—claims that some machine or other made a powered flight before the Wrights, the criterion of "flight" being in every case reduced to the level of absurdity. These claims are little more than frivolous, and have never been countenanced by any reputable historian. In this connection it is also worth mentioning that the Wright brothers' absolute priority in proper powered flight would not be affected in the least, even if their brief flights of 1903 be discounted, as no powered machines other than theirs could remain in the air for more than 20 seconds until November, 1906; and it was not until November of 1907 that a full minute's flight-duration was attained by any machine in the world except the Wrights' whose third *Flyer* had exceeded a half-hour's flight duration in 1905.

The achievements of the Wright brothers were paramount: to them, and them alone, belongs the credit of inventing, building, and flying the first successful and practical powered aeroplanes in history, and teaching the world to fly: "ils ont changé la face du monde", as Dollfus says.

Early Experiments: 1899-1901

Like Lilienthal, the Wrights were true airmen and were determined to get up into the air and fly; but they were also determined to methodically master control in the air on gliders before attempting any experiments with powered machines. They had intensified their reading and investigations, and decided to follow Octave Chanute (who himself followed Lilienthal) in adopting a biplane configuration: the general method of biplane bracing was the only technical debt they owed to Chanute.

The Wrights built their first aircraft in August, 1899: it was a biplane kite of 5 ft. span, with a fixed horizontal tailplane, designed to test the efficacy of wing warping, but with the added feature of wings which could be shifted forward or backward in relation to one another in order to control the centre of pressure (Fig. 3), the tailplane then acting as an automatic elevator.

The effect of the warping was a success, and the brothers decided to build their first full-size glider as soon as their business affairs permitted; this machine was completed by September 1900 (Fig. 4). It was a biplane of 17 ft. span and 165 sq. ft. area, but with the horizontal surface placed out front and now transformed into an elevator ("horizontal rudder"): there was warping, but no shifting forward or backward, of the wings. They believed that a front elevator would provide a safer fore-and-aft control than a rear one, especially in the event of a sudden nose-down attitude. Owing to the small wing area, this No. 1 glider was flown (in October 1900) chiefly as an unmanned kite with its controls operated from the ground: only a few manned "pilot-controlled" kite-flights were made, as well as a few piloted free glides. These tests took place over the sand-dunes of Kitty Hawk (North Carolina), a desolate location chosen as a result of their enquiries, not for its secrecy but for its steady constant winds, along with the soft sand into which crashing would be comparatively painless.

They had tried rigging this glider with a dihedral angle for automatic lateral stability, but found it flew badly and was not readily controllable in gusty conditions. So the dihedral was abandoned and seldom used by them again. There was also no inherent longitudinal stability by means of fixed horizontal surfaces.

Encouraged by their experience with No. 1, the Wrights were more determined than ever to fly; but they were still determined to proceed carefully and logically. Next year (1901) they built their No. 2 glider (Fig. 5), following the same basic configuration as No. 1, but with a wing area increased to 290 sq. ft., a span of 22 ft., a wing camber of 1 in 12, and—for the first time—an anhedral "droop" of 4 inches. The pilot lay prone—to reduce "head resistance" (drag)—in a gap in the lower wing with his hips in a cradle (swung to right or left) to operate the warping cables. They took this new glider to the Kill Devil (sand) Hills, 4 miles south of Kitty Hawk, and tested it in piloted flight during July and August 1901. Finding the camber too pronounced (with an excessive movement of the centre of pressure as the angles of incidence changed) they reduced it to 1 in 19, and so improved its performance; glides of up to 389 ft. were made, and control was maintained in winds of up to some 27 m.p.h. But the machine was far from satisfactory. The Wrights

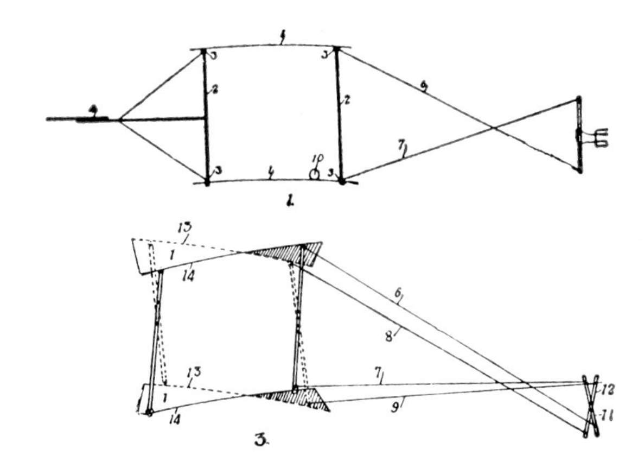

:ite with shifting wings
wing-warp controls: the
er drawing also shows
:ailplane, which auto-
cally moved up and
n (as an elevator) with
ore-and-aft shifting
ıe wings: 1899.

he No. 1 glider being
n as a kite at Kitty
/k: 1900.

he No. 2 glider, at the
Devil Hills: 1901.

began to suspect the accuracy of Lilienthal's calculations, upon which they had relied until now. The aircraft also showed an alarming tendency—when being warped—for the positively warped wings to swing *back*, and for the whole machine to slew round as it side-slipped, and crash. "Having set out with absolute faith in the existing scientific data," wrote Wilbur, "we were driven to doubt one thing after another, till finally, after two years of experiment, we cast it all aside, and decided to rely entirely upon our own investigations."

The Mastery of Glider Control: 1902

Between September 1901 and August 1902, the Wrights carried out an intensive programme of research, including the testing of aerofoil sections in a wind-tunnel and on a bicycle, as well as thoroughly re-working all their aerodynamic problems. The result was their No. 3 glider of 1902, built during August and September 1902, and tested at the Kill Devil Hills during September and October. Nearly a thousand glides were made on this machine. It was a robust biplane with a wing area of 305 sq. ft., a span of 32 ft. 1 in., and a shallow camber of 1 in 24 to 1 in 30; it had the same warping system, the same anhedral droop, and the same type of forward elevator, as the No. 2; but there was added at the rear a double fixed fin to counteract (by its weathervane action) the swing-back of the wings on the positively warped side, in the case either of regaining lateral balance after being gusted out of the horizontal, or of the pilot initiating a bank. The machine (Fig. 6), which was man-launched like its predecessor, behaved well in some of the glides; but severe trouble arose with deliberate warped banks or gust-produced (non-warped) banks, when the pilot—in trying to limit the bank or in trying to return to the horizontal—applied positive warp to the dropped wings in order to raise them: instead of coming up, these dropped wings would sink or swing back, and the machine would spin and crash.

The Wrights diagnosed the basic trouble as warp-drag (aileron-drag today) where the positive warp increased the resistance of the wings on one side, and the negative warp decreased it on the other, causing them not only to bank, but to turn about their vertical axis in the opposite direction to that originally anticipated. The addition of the fixed rear fin caused them much more trouble than it cured. For, in the acts of banking noted above, the resulting side-slip caused the fin to act as a lever, and rotate the wings about their vertical axis, thus increasing the speed (and thus the lift and height) of the raised wings, whilst retarding and lowering the dropped wings. When the pilot applied positive warp to the dropped wings, it simply aggravated the situation, and produced a spin by swinging back the dropped wings through warp-drag, thus increasing their incidence beyond the stalling angle.

This problem was solved by converting the fixed double fin into a single movable rudder, which—with its cables fastened to the warp-cradle—was always turned towards the warping direction, thus counteracting the warp-drag (Figs. 7, 8). As the rudder was also (but unintentionally) adjusted to more than compensate for the warp-drag, the machine could also be made to perform a smooth banked turn. Experience with this glider also convinced the brothers that all their machines should be made inherently unstable to allow of sensitive and immediate response to the controls.

After this vital step, the Wrights had a practical glider, with which they made some hundreds of perfectly controlled glides; they set a distance record of 622½ ft., and a duration record of 26 seconds. "The flights of 1902," wrote Orville, "demonstrated the efficiency of our system of control for both longitudinal and lateral stability. They also demonstrated that our tables of air pressure which we made in our wind tunnel would enable us to calculate in advance the performance of a machine."

;. The No. 3 glider
first version) in
light, showing the
wo fixed fins, Wilbur
piloting: 1902.

'. Launching the
No. 3 (modified)
glider, showing the
ingle rear rudder,
Orville piloting: 1902.

. The modified
No. 3 glider in
ight, Wilbur
piloting: 1902.

The First Powered Flights: 1903

Justly elated by the success of their last glider, the brothers (in March 1903) applied for a patent based on it: this was granted in 1906. They had already determined to build a powered aeroplane—they did not, as often said, put an engine into one of their gliders—and this machine was constructed during the Summer of 1903. But they had to surmount two formidable obstacles before their first Flyer (the name they gave to all their powered machines) was ready for testing, (a) the lack of an available light, yet powerful enough, engine; (b) the provision of propellers. They thereupon designed and built their own 12-h.p. motor; and—an outstanding achievement—carried out basic and original research to produce highly efficient propellers.

This first Wright Flyer (Figs. 9 to 13) was a biplane, on a skid undercarriage, of 40 ft. 4 in. span, a wing area of 510 sq. ft., and a camber of 1 in 20: it had a biplane elevator out front, and a double rudder behind, whose control cables were linked to the warp-cradle: the motor drove two geared-down pusher propellers through a cycle-chain transmission in tubes, one being crossed to produce counter-rotation. The launching technique was as follows: the Flyer's skids were laid on a yoke which could run freely on two small tandem wheels along a 60-foot sectioned wooden rail, laid down into wind; the machine was tethered whilst the engine was run up, and then unleashed; when its speed produced sufficient lift, it rose from the yoke and flew. The Wrights did not use any accelerated take-off device for their 1903 flights. The weight (empty) of the Flyer was 605 lbs.

After minor but exasperating set-backs at the Kill Devil Hills—where the tests took place—and after brushing up their piloting on the 1902 glider, the first attempt was made on December 14th, 1903, with Wilbur at the controls (he had won the toss of a coin): but owing to over-correction with the elevator, the Flyer ploughed into the sand immediately after take-off (Fig. 12).

It was on the morning of Thursday, December 17th, 1903, between 10.30 a.m. and noon, that the first flights were made. After five local witnesses had arrived, Orville (whose turn it now was) took off at 10.35 into a 20–22 m.p.h. wind and flew for 12 seconds, covering 120 feet of ground, and over 500 feet in air distance (Fig. 13). On the fourth and last flight, at noon, Wilbur flew for 59 seconds, covering 852 feet, and over half a mile in air distance. Their speed was about 30 m.p.h. All four take-offs on December 17th were made from flat ground. These flights were the first in the history of the world in which a piloted machine had taken off under its own power; had made powered, controlled, and sustained flights; and had landed on ground as high as that from which it had taken off. No aeroplanes other than the Wrights' could remain in the air for more than 20 seconds until November 1906; it was not until November 1907 that a full minute's duration was achieved by a European machine.

This historic first Flyer—following a most lamentable attack on the Wrights—was loaned to the Science Museum, London, from 1928 to 1948. It is now preserved in the National Air and Space Museum (Smithsonian Institution) at Washington, D.C., a reproduction being shown in London.

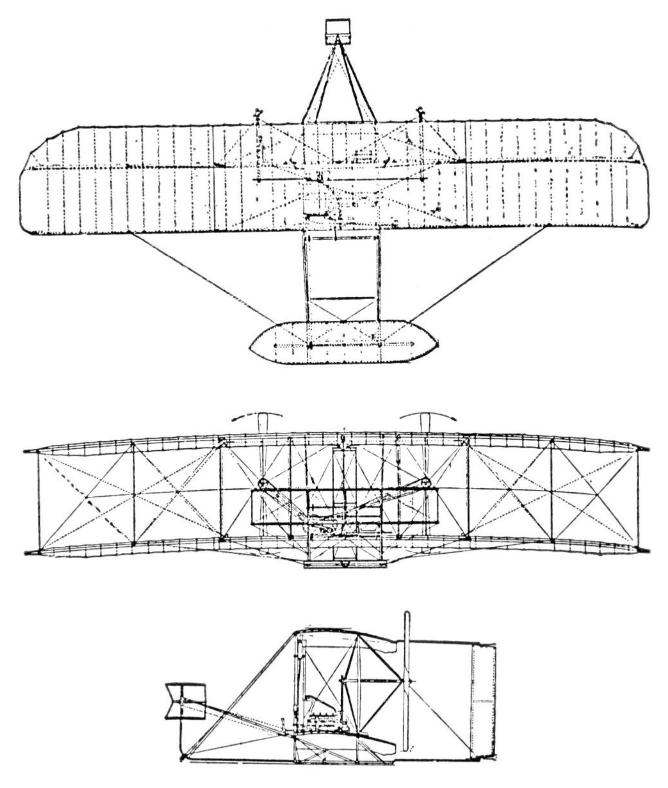

9. General arrangement drawings of the Wrights' first powered machine (Flyer I): 1903.

10. The original Wright Flyer I, when on loan to the Science Museum, London.

11. Side view of the Wright Flyer I, at the Kill Devil Hills: 1903.

12. The first attempt at flight (Wilbur piloting): 14th December, 1903.

13. The world's first powered, sustained, and controlled flight (Orville piloting): 17th December, 1903.

Improved Power-Flying: 1904

14. The Wright Flyer II in flight at the Huffman Prairie: 1904.

The Wright Flyer II was completed in May 1904, and through the kindness of a friend, an "aerodrome" was set up at the Huffman Prairie, a 90-acre pasture at Simms Station, about 8 miles east of Dayton. The new Flyer had approximately the same dimensions as the first, but had less camber (1 in 20 to 1 in 25), and a new engine of 15–16 h.p.: the pilot still lay prone, and the warp and rudder controls were still linked (Fig. 14). From May 23rd to December 9th, about 80 short flights were made which enabled the brothers to obtain practice in controlling and manoeuvring a powered machine. Some 100 starts were made in all, and various minor setbacks had to be overcome before consistent and productive flights were achieved. Their total airborne time was about 45 minutes: the longest flight lasted for 5 min. 4 sec. and covered about $2\frac{3}{4}$ miles. On September 7th they introduced for the first time their weight-and-derrick assisted take-off device (see pages 20, 21), to make them independent of the weather, in view of the small area of the pasture. The most important event was their first circuit (by Wilbur on September 20th), which later became a necessary commonplace, as they did not want to overfly other property: this first circuit was the subject of a detailed eye-witness report made and published by Amos I. Root—the first such report in history of a powered aeroplane flight (see pages 34, 35).

There was one control problem still outstanding, a tendency to stall in tight turns (see page 12), which was not to be solved until 1905.

It was early in this 1904 season that two press visits took place; and, had not the engine failed on both occasions, the history of aviation—and indeed of civilization itself—would have been greatly changed: the reporters never came back, despite every resident in the neighbourhood reporting the numerous flights, and soon taking them for granted.

The airframe of the 1904 Flyer II was broken up and destroyed in 1905.

The First Practical Aeroplane: 1905

Although the four brief flights of 1903 have naturally invested the Wrights' first Flyer with the greatest fame, their Flyer III of 1905—seldom dealt with in aviation histories—should stand equally with it; for the 1905 machine was the first practical powered aeroplane of history (Figs. 15, 16). It was of the same general arrangement as the others, but noticeable differences appeared in the placing of the elevator farther forward and the rudder farther back to improve longitudinal control. The span was 40 ft. 6 in.; the wing area was slightly reduced, to 503 sq. ft.; the camber was increased to 1 in 20; new sets of propellers were used; but the excellent 1904 engine was retained. The prone pilot position was retained, and also—for the start of the season—the warp and rudder linkage. Its speed was approximately 35 m.p.h. Like all the Wright aircraft, it was built inherently unstable and had to be "flown" all the time by the pilot. The rudder outrigger was sprung to allow it to hinge upwards if it dragged on the launching rail, or the ground.

The 1905 season at the Huffman Prairie lasted from June 23rd to October 16th, over 40 flights being made. But now the Wrights were concerned with reliability and endurance: they were airborne this season for just over 3 hours. In September, the trouble they were having in tight turns was diagnosed as a tendency of the lowered wings to slow up and stall; and the cure seen to be in putting down the nose to gain speed whilst turning. It was in seeking this cause and cure that the brothers took the important step of unlinking the warp and rudder controls, and providing for their separate, or combined, operation in any desired degree.

With this Flyer now perfected, the Wrights made many excellent flights, including durations of 18 min. 9 sec., 19 min. 55 sec., 17 min. 15 sec., 25 min. 5 sec., 33 min. 17 sec., and—on October 5th—their record of 38 min. 3 sec., during which they covered over 24 miles.

The description of this machine as the world's first practical powered aeroplane is justified by the sturdiness of its structure, which withstood constant take-offs and landings; its ability to bank, turn, circle and perform figures of eight; and its reliability in remaining airborne (with no trouble) for over half an hour. It is now preserved in a specially built hall in Carillon Park at Dayton (Ohio).

It was in 1905 that the Wrights first offered their invention to the U.S. and British governments. In January, on the experience they had had with the Flyer II, an offer was made to the U.S. War Department, which was refused outright without even an attempt at investigation—"a flat turn down", as the exasperated Wilbur called it. They then offered it to the British War Office. The British dilatoriness was as bad as the U.S. refusal; so, with persuasion from Chanute, the brothers again (in October, still of 1905) offered it to the U.S. War Department, with all the added confidence now provided by the performance of their Flyer III. Again it was turned down; again the War Department assumed the Wrights were asking for financial assistance—whereas they made it quite clear they were offering a finished product—and, this second time, the Department made one of the classic statements of aviation history: "the device must have been brought to the stage of practical operation without expense to the United States".

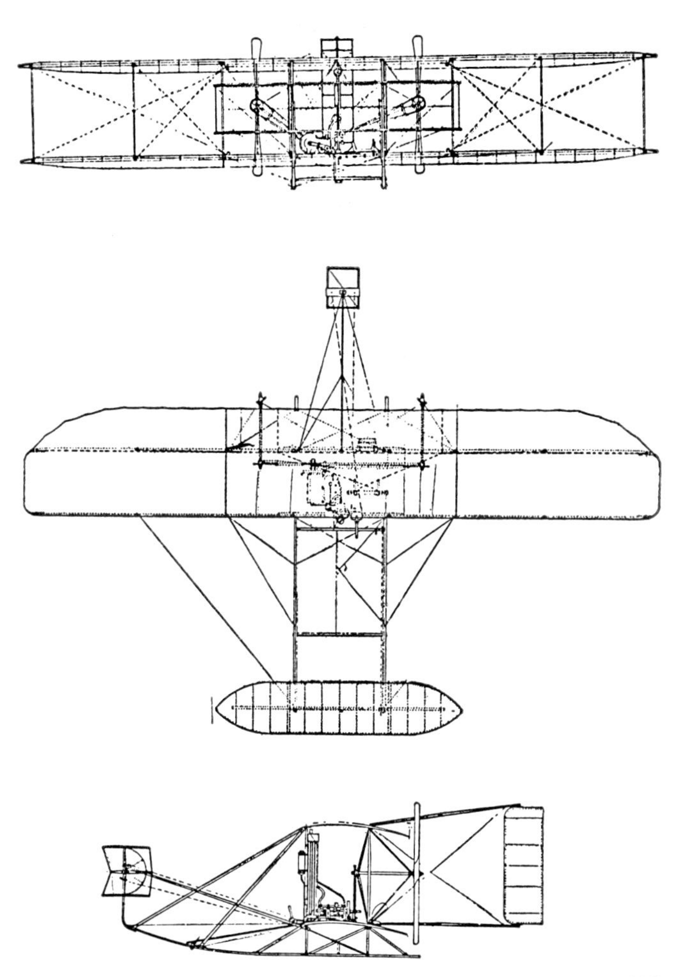

15. General arrangement drawings of the Wright Flyer III: 1905.

16. The Flyer III in flight at the Huffman Prairie: 1905.

The Patent and the Flight Control Systems

The basic Wright patent in the United States (No. 821,393) was applied for in 1903 and granted in 1906. The patent described and illustrated (Fig. 17) what was basically the final form of the 1902 glider; the provisions covered the principle of increasing the angle of incidence at one wing tip, and simultaneously decreasing it at the other; the practical technique of wing-twisting (the so-called "warping"); and, most important of all, the simultaneous use of warping and rudder to effect proper lateral control.

An important modification of the helicoidal twist method of warping the wings overall was permanently adopted with the first powered Flyer of 1903, in which warping was confined to the two outer bays on each side, where the trailing edges of both sets of wings could be lowered or raised (see below).

The developed Wright control system, as fitted to machines of what we may call their standard type A (from 1908 onwards), was operated entirely by two hand-levers: there was no mechanism worked by the feet.

Oddly enough—where the warp and rudder controls were concerned—there came to be a "Wilbur" sub-system of levers and cables (used in France, Italy, and England), and an "Orville" sub-system which, after being used in Germany, became the standard arrangement in the U.S.A.: the elevator lever (on the left) remained the same in both (A in Fig. 18).

In the "Wilbur" sub-system (Fig. 18), the right-hand lever (B) was rocked laterally to warp the wings, left to bank left, right to bank right, *via* the cables DD; this lever was also moved forwards or backwards to operate the rudder (E) *via* a pivoted bar (C) and crossed cables (CC), forwards to put on left rudder and backwards to put on right: any desired combinations of warp and rudder were obtained by moving the lever obliquely. F is the seat (all standard machines had two); G is the fixed foot-rest.

The "Orville" sub-system is seen from the front in Fig. 19. The lever (A) could only be moved forwards or backwards, and this motion operated the wing-warping by means of a semi-disc (B) fixed to the base of the lever, *via* two cables (CC), and also put on a small amount of rudder automatically, unless corrected by the handle (D): the rudder was operated by this handle (D) at the top of the lever, which was rocked laterally (to the left for left rudder, to the right for right), *via* a bell-crank lever connected by a rod to a disc (E) rotating freely on the lever shaft, from which two crossed cables (FF) went aft to the rudder. Combinations of warp and rudder were obtained by simultaneous movements of the lever and its rocking handle.

The wing-warping was effected by two cable systems (Fig. 20). The active system (AA, etc.) pulled down the two outer rear struts—and hence part of the wing tip—on one side or the other; whilst the passive system (BB, etc.) automatically pushed up the opposite two struts (and wing tip), thus giving a helicoidal twist to the wing-ends. Wing-warping in practice is shown in Fig. 21, where a British-built Wright glider of 1909 is seen just as the wings were being warped, and a second or two before the warp had taken effect and banked the machine to starboard.

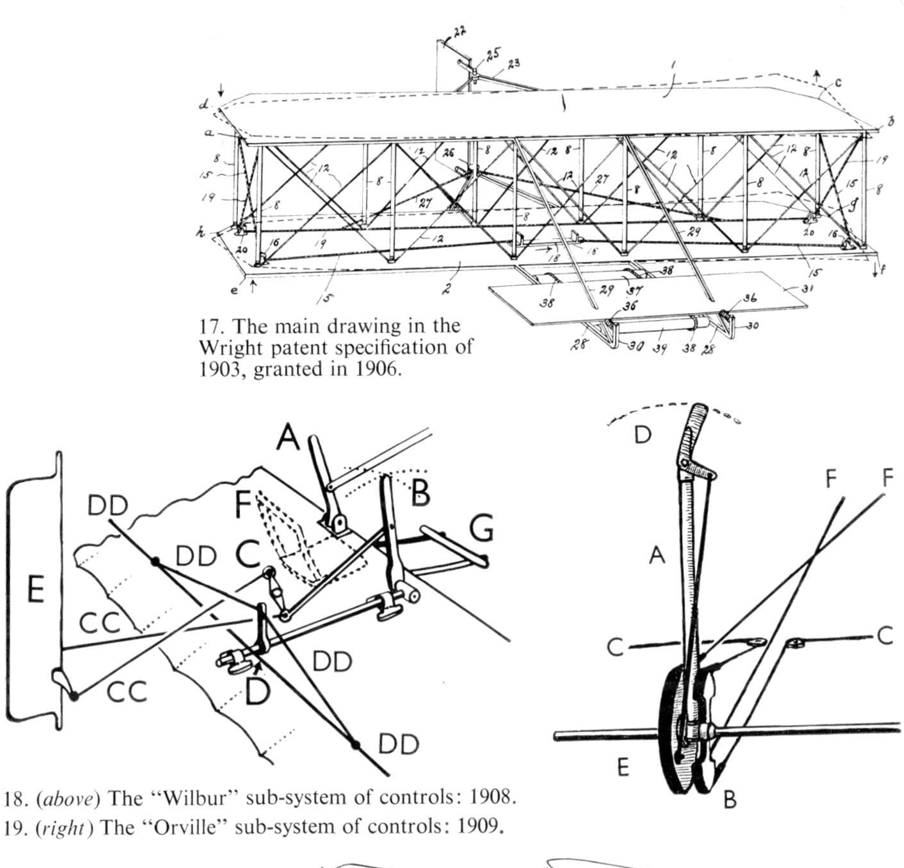

17. The main drawing in the Wright patent specification of 1903, granted in 1906.

18. (*above*) The "Wilbur" sub-system of controls: 1908.

19. (*right*) The "Orville" sub-system of controls: 1909.

20. The two cable systems to effect the wing-warping: the lines (AA) represent the active system, the lines (BB) the passive system.

21. Wing-warping seen in action, on a British-built Wright glider: 1909.

The standard Wright type A: 1908-9

The Wrights did not once leave the ground between October 16th, 1905 and May 6th, 1908—a period of 2½ years—nor did they allow anyone to view their machine. This astonishing interregnum, which severely retarded the whole development of flying, was due basically to the continued thwarting of the Wrights' legitimate demand that any client must guarantee to purchase their machine provided they (the clients) agreed the desired performance, and provided that the machine performed as agreed. But the clients, particularly the U.S. Government, insisted on viewing the machine (or drawings of it) before signing a contract, an unreasonable and totally unacceptable condition to the Wrights, especially in view of the many would-be spies who had heard reports of the 1904 and 1905 flying, and were out to learn all they could.

But this 2½ years' interregnum was not wasted technically, because the brothers built some half-dozen improved engines, and two or three new Flyers, pending a satisfactory agreement with their own or a foreign Government, or—failing that—some commercial firm. These machines (Figs. 22, 23) may now be called collectively the Wright type A; these include the machines built in 1907 (first used in 1908) and later, along with those built under licence in France, Britain, and Germany in 1909; the Wrights themselves built about seven of these standard machines during the period 1907–09, as well as the somewhat modified one built for the U.S. military trials in 1909 (see page 24): six were built under licence by Shorts in England, but the numbers built on the Continent are not yet known. All the machines were similar, with only minor differences—such as the Wilbur and Orville "sub-systems" of warp and rudder control—and represent not only the culmination of the Wrights' achievement, but the type of Wright machine which was first seen in public, and which directly inspired the last and triumphant phase of world aviation in which the powered aeroplane was established as a new and practical vehicle in the year 1909.

The approximate specification of the type A was:

Wing area: about 510 sq. ft.	Span: 41 ft.
Chord: 6 ft. 6 in.	Gap between wings: 6 ft.
Elevator area: 70 sq. ft.	Rudder area: 23 sq. ft.
Length: 31 ft.	Engine: 4 cyl., 30–40 h.p.
Weight (empty): 800 lb.	Speed: 35–40 m.p.h.

This type, similar in general configuration to the 1905 Flyer III, was however, a two-seater, with upright seating: it still retained the skid undercarriage and derrick-and-rail launching, although it could (and occasionally did) take off from the rail on engine power alone. Also, as before, it was inherently unstable, and it was this feature which particularly struck the Europeans as undesirable in an ordinary "workaday" aeroplane.

It is extraordinary, and historically disgraceful, to find that only one example of this classic type of aeroplane now survives; and—considering it is one of aviation history's half-dozen most important relics—that it is comparatively unknown: it is Orville's Berlin machine of 1909, the Wright A (Berlin), now preserved in the Deutsches Museum at Munich.

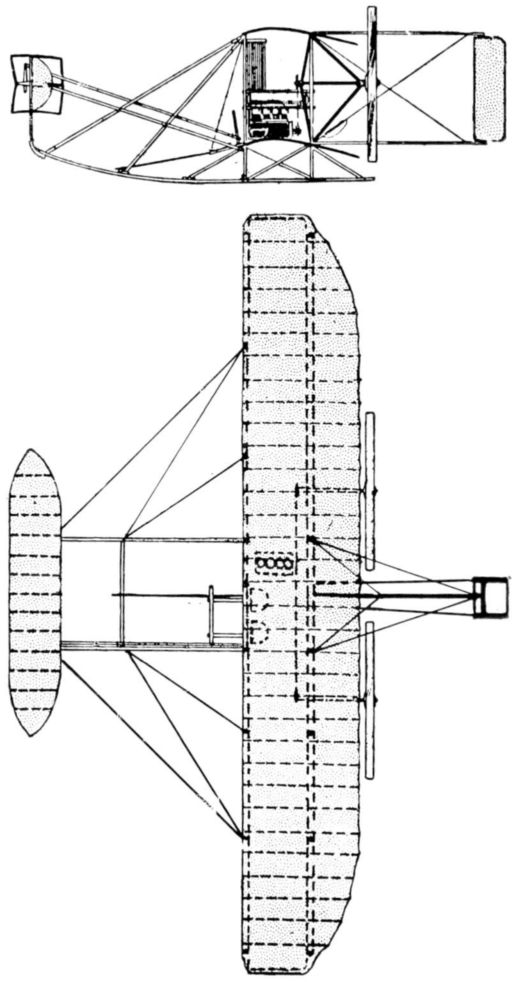

22. General arrangement drawings of the standard Wright type A: 1908–9.

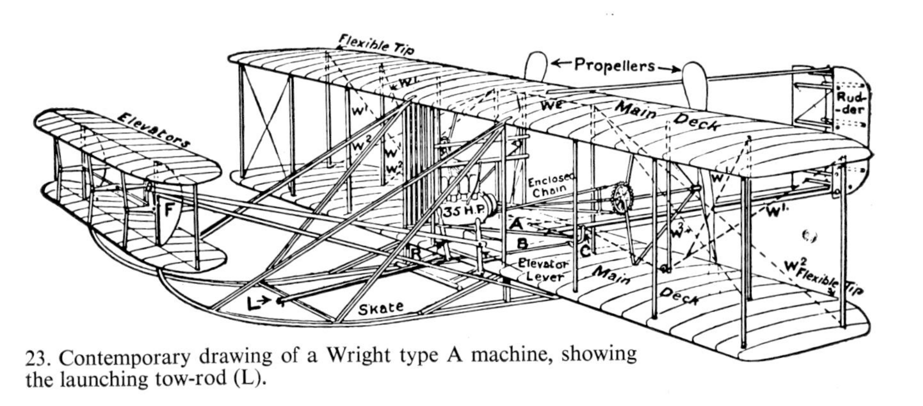

23. Contemporary drawing of a Wright type A machine, showing the launching tow-rod (L).

Wilbur Wright in France: 1908

At last, after interminable negotiations, the Wrights signed a contract with the U.S. Army in February 1908, and with a French company the following month. It was decided that Orville would conduct the Army tests, and Wilbur demonstrate in Europe, where (at Le Havre) a Flyer already lay crated, having been sent over to France in July of 1907.

To regain their skill, the brothers took the 1905 Flyer III, now adapted to take two, both seated upright, to the Kill Devil Hills, and practised there. From May 6th to 14th, 1908, they made some 20 flights, among which were the world's first two passenger flights, when each brother took up C. W. Furnas on 14th May, the best of these latter lasting for 3 min. 40 sec.

Wilbur went to France late in the same month, and—after various delays—completed the assembly of the 1907-built Flyer in his friend Leon Bollée's factory at Le Mans. Intense and widespread interest, suspicion, and scepticism, was focused on Wilbur by both the French aviators and the Press, following the reports of the Wrights' power-flying of 1904 and 1905. Therefore, when Wilbur announced his first flight in public—he had so much confidence in the machine, that he made no secret flight-tests—the most critical of technical audiences assembled on August 8th, 1908 at the small local racecourse at Hunaudières (5 miles south of Le Mans).

On that memorable day, Wilbur took off, made two graceful circuits, and landed smoothly: he was in the air for only 1 min. 45 sec.; but this machine in Count de La Vaulx' words, "revolutionised the aviators' world."

European flying, despite its devotees, was still in a parlous condition, its worst feature being the lack of lateral control and proper manoeuvrability in the air—the first European circle had only been made that January—to say nothing of inefficient airscrews and other troubles. It is hard for us today to realise the shock of almost stunned amazement which greeted Wilbur's effortless mastery of control, his graceful banks and turns. "No one can estimate," said one French spokesman, "the consequences which will result from this new method of locomotion, the dazzling beginnings of which we salute today." "For us in France," exclaimed Blériot, "and everywhere, a new era in mechanical flight has commenced . . . it is marvellous." "It is a revelation in aeroplane work," said René Gasnier, "who can now doubt that the Wrights have done all they claimed? . . . We are as children compared with the Wrights." "Mr Wright has us all in his hands," said Paul Zens. The French aviation press was equally repentant and enthusiastic: "one of the most exciting spectacles ever presented in the history of applied science." And finally, here is the great French pioneer, L. Delagrange: "Well, we are beaten. We just don't exist." ("Eh bien! Nous sommes battus! Nous n'existons pas!")

Between August 8th and 13th Wilbur made nine flights, the longest lasting just over 8 minutes (Figs. 24–27). Then he received permission to use a great military ground, the Camp d'Auvours, 7 miles east of Le Mans. Here, from August 21st to the last day of December 1908, he made 104 flights, and was airborne for about 25½ hours, thus making some 26 hours for the combined French locations that year.

[Continued on page 20.]

:4. Wilbur flying
his Wright A
(France) at
Hunaudières:
August, 1908.

:5. Wilbur coming
n to land at
Hunaudières:
August, 1908.

6. The machine on
s handling dollies
utside the hangar
t Hunaudières:
August, 1908.

7. Close-up (from
rear) of the machine,
owing early-type
ropellers: August,
908.

Wilbur's astonishing 1908 "season" at Auvours included; taking up passengers on some sixty occasions; fourteen flights of between $\frac{1}{4}$ and $\frac{1}{2}$ hour duration; six of between $\frac{1}{2}$ and $\frac{3}{4}$ hour; six of between 1 and 2 hours; and his record (on December 31st) of 2 hours 20 minutes 23 seconds. Also one flight to gain the altitude record of 360 feet (on December 18th).

The world—not only of aviation—soon endorsed the now famous statement of Major B. F. S. Baden-Powell (the then Secretary of today's Royal Aeronautical Society): "that Wilbur Wright is in possession of a power which controls the fate of nations, is beyond dispute".

The Camp d'Auvours soon became a Mecca for all those seriously interested in aviation. As Wilbur went from strength to strength, his achievements, as well as his compellingly modest personality, won him world renown.

One of the first admissions to be made by even the most sceptical of Frenchmen was that, to have advanced so far, the Wrights' claims to having flown from 1903 to 1905 were obviously true.

The technical aspects of the Wright machine were analysed in minute detail by the French and British visitors; the accent was always on the primitive flight-control in Europe, and its superb mastery by the Wrights. But the Europeans, although they quickly realised the sensitivity to its controls of the deliberately unstable Wright machine, decided that a good measure of automatic stability should be built into future aircraft. They also came to wonder why the Wrights held to their rail and derrick take-off technique, when aerodromes were rapidly being prepared all over Europe.

As this technique is of great historical interest, a description of it is given here. The Wright machine had to be wheeled about on two handling dollies lashed to the lower wings: they are seen in position in Fig. 26, and removed on the left of both Fig. 29 and Fig. 30. For launching, the sectioned wooden rail ("monorail"), 100 feet long, was laid down into wind; and the derrick, with its 15$\frac{3}{4}$ cwt. (800 kg). weight, was set up behind it (Figs. 29, 30). The weight was raised either by multiple man-power (Fig. 29), or by an automobile. The $\frac{1}{2}$ inch rope ran from the weight up to, and over, pulleys in the top of the derrick; down to the base of the derrick, and round under another pulley; along beside the rail; up and around a final pulley at the rail-end; and then back to the aircraft: here, a metal eye in the rope's end was slipped over a pin projecting *downwards* from the launch tow-rod of the aircraft, which was hinged to the lower wing (see letter L in Fig. 23). The aircraft's skids rested freely on a yoke ("truck") running on tandem rollers. When the weight was released, the machine was pulled rapidly along the rail: as it was nearing the end, the pilot raised the nose, whereupon a cross-bar of the elevator-outrigger met and stopped the tow-rod, causing the eye of the rope to slide off the pin, and the machine took off (Fig. 31). It was a laborious, but extremely reliable, technique of launching an aeroplane: it never failed in thousands of take-offs. An interesting detail was still—as noted of the 1905 Flyer III—the spring device which allowed the rudder-outrigger to hinge upwards if the rudder hit the ground. Despite the obvious advantages of the wheeled undercarriage, the Wrights did not adopt wheels until 1910, although individual owners sometimes added wheels to the skids of their standard Wright machines (see Fig. 40), the first being L. Schreck in France in September 1909.

28. Wilbur's first woman passenger, Madame Hart O. Berg: note the "bent end" propellers, and the string round the lady's skirts to avoid both aerodynamic and moral hazards: October 7th, 1908.

29. Volunteers hauling up the aunching weight at Auvours: August or September, 1908.

0. The engine being started, rior to the machine being aunched from its rail at uvours: August or Septem- er, 1908.

1. Wilbur taking off from the il at Auvours: August or eptember, 1908.

Orville Wright in America: 1908

When Orville Wright embarked on the acceptance tests for the U.S. Army in September 1908, he flew as spectacularly as his brother in France (Figs. 32–34): the tragedy that ended these trials did nothing to diminish his triumph. Wilbur wrote to their sister: "Tell Bubbo (Orville) that his flights have revolutionised the world's beliefs regarding the practicability of flight. Even such conservative papers as the London *Times* devote leading editorials to his work and accept human flight as a thing to be regarded as a normal feature of the world's future life."

Supervised by officers of the U.S. Signal Corps, Orville started the tests at Fort Myer (near Washington, D.C.) on September 3rd 1908, and crashed—killing his passenger Lieut. T. E. Selfridge—on September 17th (Fig. 35). But during those few days he made ten flights, and was airborne for just under six hours: the flights included one of 57 min. 25 sec., four of over an hour each (1 hr. 2 min. 15 sec.; 1 hr. 5 min. 52 sec.; 1 hr. 10 min. 24 sec.; 1 hr. 14 min. 20 sec.) and three passenger flights. During two flights he created altitude records of 200, then 310 feet, surpassed by Wilbur on December 18th.

Although the first official flight in public in the United States had been made on July 4th, 1908, when Glenn Curtiss flew some 6,000 feet in 1 min. 43 sec., this was a primitive achievement ranking with similar flights in Europe; and it was Orville's flying in September that brought home to the American public that the air had really been conquered.

The fatal crash—the first in powered aviation—happened at 5 p.m. on September 17th. The machine was of the standard Wright type A, with "Orville" controls. Orville had taken up Lieut. Selfridge as an official passenger, and was making the fourth round of the field when the trouble struck. A blade of the starboard propeller developed a longitudinal crack which caused it to flatten and lose its thrust, thereby setting up an imbalance with the good blade: the consequent violent vibration loosened the supports of the propeller's long shaft, causing the latter to "wave" to and fro, and thus enlarge the propeller disc: the good blade then hit, and tore loose, one of the four wires bracing the rudder outrigger to the wings, the wire winding itself round the blade and breaking it (the blade) off. Orville cut the motor, and tried to land; but the rudder canted over and sent the machine out of control: even then, in the ensuing nose-dive, Orville was succeeding in bringing up the nose with the elevator; but it was too slow, and the Flyer crashed, killing Selfridge. Orville was seriously injured, but made a good recovery.

From Lilienthal to the end of 1908—that is to say during the birth years of the practical aeroplane—only four men had been killed, the first three on gliders; Lilienthal (1896), Pilcher (1899), Maloney (1905), and Selfridge (1908); a very small total considering the risks.

The death of Selfridge did nothing to deter the United States Army's interest: having seen the outstanding performances Orville put up, they were determined to go ahead as soon as Orville had recovered.

32. Orville's Wright A (Fort Myer), at Fort Myer, U.S.A. (note the "bent end" propellers): September, 1908.

33. Orville taking Maj. G. O. Squier for a flight at Fort Myer: September 12th, 1908.

34. Orville flying alone at Fort Myer: September 12th, 1908.

35. The fatal crash at Fort Myer, when Lieut. Selfridge was killed and Orville injured: September 17th, 1908.

The Wrights in 1909

In 1909 the Wright Brothers reaped the universal honour and glory—official and popular—due to them. This was the year which not only saw the culmination of their own pioneer flying; but, as a direct result of their displays in 1908, the arrival of the European aeroplane as a practical and accepted vehicle, symbolised by Blériot's Channel crossing on July 25th and the Reims meeting in August. As soon as the Europeans had grasped the full implications of the Wrights' lateral control—Blériot used the Wrights' wing-warping on his cross-Channel machine—they built an excellent group of biplanes and monoplanes by mid-1909.

In January 1909, Wilbur Wright moved to the warmer climate of Pau (in South-West France) where he flew from February 3rd to March 20th (Fig. 36), his chief concern being the completion of the training—started at Auvours—of the three French pilots, called for in his contract. He was joined there in January by the convalescing Orville and their sister Katharine.

On April 1st they arrived in Rome; and from April 15th to 25th Wilbur gave demonstrations and passenger flights to great acclaim at Centocelle (near Rome); started the training of two Italian lieutenants; and took up a cinematographer. His aircraft was a new, but still standard, Dayton-built machine recently assembled at Pau.

At the end of April, the Wrights started their slow triumphal return home to Dayton, being honoured and fêted en route at Paris, Le Mans, London, New York and Washington.

On June 28th Orville made the first of his "warming up" flights at Fort Myer, in preparation for his new Army trials. The aircraft was a "one-off" modification of the standard type A, specially built for the trials, with reduced wing area (415 sq. ft.), higher speed (about 45 m.p.h.), a higher skid undercarriage, and with no sprung rudder outrigger (Fig. 37). By July 30th, after some typically excellent flying, the official tests were successfully completed, and the machine was bought by the Army. It is now preserved in the National Air and Space Museum at Washington, D.C. and known as the "Signal Corps" machine.

In France, the first great aviation meeting was held at Reims (August 22nd to 29th), in which three Dayton-built Wright machines flew, one of the most popular Wright pilots being Eugène Lefebvre (Fig. 38).

In August, Orville went to Germany, and between August 29th and October 15th made many successful exhibition and training flights at Tempelhof, near Berlin (Fig. 39) and Potsdam. His machine (built in America) is now preserved in the Deutsches Museum at Munich, and is the only standard type A Wright aircraft to survive.

Meanwhile, back in America, Wilbur was making demonstration flights in September and October; and from October 8th to November 5th he instructed the first three Army pilots at College Park, Maryland, where he also took up America's first woman passenger (Mrs. Vanderman) on October 27th.

The first pilot to receive a pilot's certificate (brevet)—No. 8—on a Wright machine was Wilbur's pupil, the expatriate Russian—of French descent—Count Charles de Lambert. The first Wright pilot to be killed was Lefebvre, who crashed—cause unknown—at Juvisy on September 7th 1909.

6. Wilbur taking a
assenger at Pau:
ebruary or March,
909.

7. Orville's Wright
. (Signal Corps)
achine at Fort
Iyer for the second
rmy trials: July, 1909.

3. Eugène Lefebvre
ying his Wright A
achine at the Reims
eeting: August 1909.

9. Wheeling out
rville's Wright A
Berlin) machine at
empelhof, Berlin:
ptember, 1909.

Transition and Consolidation: 1909-10

After the new surge of activity in 1909, aviation rapidly became an established industry, drawing many technicians—and also pilots—from the sphere of the automobile. European aviation may be said to have come of age by the end of 1909; and although the Wrights were universally acknowledged as their inspirers and mentors, the Europeans soon came to regard the Wright machines as simply one type of successful biplane, with the disadvantages of its inherent instability, and its skid undercarriage.

In the United States, the Wright machines were to remain paramount for some time, but they soon had a serious rival (from 1909 onwards) in the Curtiss biplanes.

In Europe, the first important modification to the standard type A Wright machine was the "unofficial" addition of four small wheels to the skids (Fig. 40): this was first done in September 1909 in France, by L. Schreck.

The next important modification to the standard machine came in 1910, when a fixed tailplane was added behind the rudder, supported by a separate auxiliary outrigger built around the existing two-boom rudder outrigger (Fig. 41): this tailplane was soon altered to become a second elevator, working in concert with the front elevator. It was in a French-built machine of this latter type—closely similar to that shown in Fig. 41, but fitted also with wheels on the skids—that the Hon. C. S. Rolls was killed at the Bournemouth meeting on July 12, 1910.

Also in 1910—in the U.S.A.—the old twin-boom outrigger was exchanged for a frame of four longerons joined at the rear end by two tall uprights and two short horizontals, with the rudder enclosed within this structure, and the rear elevator hinged (behind) to the uprights, and working in concert with the front elevator (Fig. 42).

"The difficulty in handling our machine," wrote Orville to Wilbur in September 1909, "is due to the rudder (i.e. elevator) being in front, which makes it hard to keep on a level course. . . . I do not think it is necessary to lengthen the machine, but to simply put the rudder behind instead of before."

It was thus in 1910 that the Wrights produced their first machine with no forward elevator, and the first with a wheeled undercarriage: this was the so-called "Model B". It was in effect a further, but drastic, modification of the type A, with a span of 39 ft. (possibly 38 ft. 6 in. on some) and a wing area of about 500 sq. feet: as in the last modification of the standard type, the rudder was enclosed in the outrigger structure, and the elevator placed out back: in addition to the wheeled undercarriage, two triangular "blinkers" were fixed on the forward skids to augment the keel-area. The photograph which best shows these characteristics is of a modified Model EX (Fig. 43) which was a smaller single-seat version of the Model B. This particular machine—the *Vin Fiz*—made a celebrated flight from Long Island to Long Beach (California) in 1911. The *Vin Fiz* is in the National Air and Space Museum, Washington, and a Model B is preserved in the Franklin Institute, Philadelphia.

40. The French-built Wright A (Baratoux), with wheels added to the skids: 1909.

41. A French-built type A Wright machine, with an additional outrigger bearing a fixed tailplane: 1910.

42. A modified Wright type A, with a 4-boom tail outrigger bearing an auxiliary elevator: 1910.

43. A modified Model EX Wright biplane: 1911.

End of an Epoch: 1910-11

This brief survey of the Wrights' work is intended mainly to show how they conquered the air, and inaugurated what may still be called the Air Age. Wilbur was to die of typhoid fever in 1912; but Orville was to live to the ripe age of 77, dying in 1948. Orville gave up piloting in 1918, but continued for many years in productive and inventive aeronautical work.

It is often asked whether the brothers ever broke their rule of not flying together: they did, once, on May 25th, 1910, when Orville took Wilbur for a short flight. On the same day, Orville also took their 82-year old father, the Bishop, for the only flight of his life.

I am closing this survey—apart from a supplementary section on the influence of the Wrights—with Orville's gliding in 1911; here the story came full circle after a decade: it was in 1901 (with their second glider) that they first made proper glides at the Kill Devil Hills: now, in 1911, Orville returned there to glide once more.

But first there is an important machine and it variants to mention, which precedes the glider. It is the Model R of 1910, also called the "Roadster" and "Baby Wright" (Fig. 44): this was a single-seat competition machine, with a span of 26 ft. 6 in., a wing area of 180 sq. ft., a 30–35 h.p. engine, and a speed of 50 m.p.h. A smaller version of it—the "Baby Grand" was flown by Orville at the Belmont Park meeting in 1910, where he attained a speed of 70–80 m.p.h. (Fig. 45).

In October 1911, Orville and the English pioneer Alec Ogilvie, flew a new Wright glider at the Kill Devil Hills. Orville had also intended to test an automatic stabiliser he had been developing, but this was not done. The machine was a biplane of 32 feet span, and a wing area of 300 sq. ft. A fixed vertical fin was placed forward, at first attached to a front strut (Fig. 46*a*), then set 5 feet out in front (Fig. 46*b*). Many successful soaring flights were made, including a world record (by Orville) on October 24th—9 min. 45 sec.—which was to stand for a decade.

The story of Wilbur and Orville Wright presents one of the most quietly dramatic epics of history: to a greater degree than any other invention of modern times, the aeroplane has transformed our world and our lives, partly for good, partly for ill. The absorbing interest of the events often obscures the superlative qualities of mind and spirit possessed by both the Wrights, which alone made possible their triumphs. Wilbur, the elder of the brothers, was the first to be inspired by the idea of human flight, and the following entry in his father's diary will, I feel, provide a fitting epitaph for both men, and a fitting epilogue to this short account of their work:

> "This morning, at 3:15, Wilbur passed away, aged 45 years, 1 month and 14 days. A short life full of consequences. An unfailing intellect, imperturbable temper, great self-reliance and as great modesty, seeing the right clearly, pursuing it steadily, he lived and died."

4. A Model R
"Baby Wright")
iplane: 1910.

5. Orville on his
Baby Grand" at
elmont Park: this is
tted with the only
Vright V-8 engine to
e flown: 1910.

5A. Orville's new
ider at the Kill
evil Hills (with the
n on a front strut):
ctober, 1911.

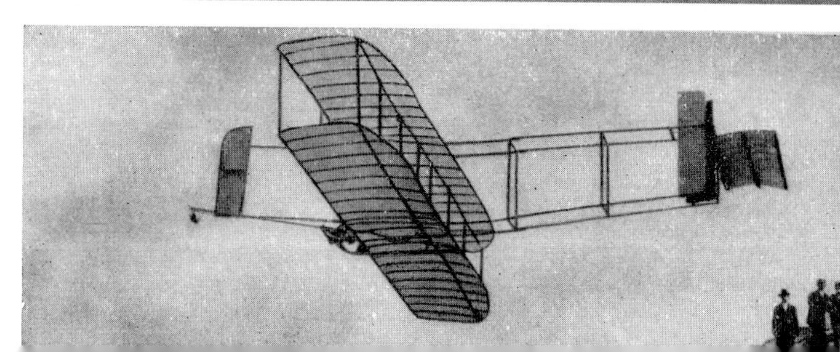

5B. The same glider,
ith fin on an out-
gger (this machine is
ten mistaken for the
02 glider): 1911.

The Influence of the Wright Brothers: 1902-9

Aviation in Europe had been virtually at a standstill since Lilienthal's death in 1896. It was information and illustrations of the Wrights' gliders, then news of their powered flights, which revived European aviation from 1902 onwards.

The French Captain F. Ferber was the first European (in 1902) to seek and obtain (from Chanute) information about the Wrights' gliding in 1900 and 1901. He thereupon abandoned his unsuccessful attempts to imitate Lilienthal, and in 1902 built a poor imitation of the 1901 Wright No. 2 glider (Fig. 47), which next year he tried to improve, but without success.

In April 1903 Octave Chanute, the Wrights' friend, lectured in Paris on the brothers' 1902 glider; both his words and the photographs he showed were also widely published. The effect of this news was to directly inspire the whole French aviation movement to action. However, although both the warping and the simultaneous use of warp and rudder were mentioned, the French took little notice of the implications and did not persevere with gliding, although various imitations of the 1902 glider were made, such as Ernest Archdeacon's in 1904 (Fig. 48). One of the main reasons for this failure to progress was an article by Robert Esnault-Pelterie in which he informed the world that in 1904 he had built an exact replica of the Wright 1902 glider—which he had not—and that it was a failure. Incidentally, having abandoned warping, he was the first to apply—but not invent—ailerons, which were actually elevons; but they, too, were unsuccessful (Fig. 49). Other serious hindrances to the Europeans' development included the lack of research on propellers—which remained primitive affairs until 1909—and the inexplicably dilatory and slap-dash methods of progression.

The effect of this, and other actions, was threefold; (1) it retarded the Europeans and sent them off in a variety of fruitless directions; (2) it led to neglect of control in roll; (3) it intensified the attempts at automatic stability, especially longitudinal, Ferber's 1904 machine (Fig. 50) being the first.

In 1905 appeared the tentative Voisin-Archdeacon float-glider (Fig. 51)—towed by a motor-boat—which added the Hargrave box-kite to European aviation. In 1906 came Santos-Dumont's famous but sterile "14-*bis*" (Fig. 52) which made the first European powered hop-flights, the best of which covered 220 metres (say 722 ft.) in $21\frac{1}{5}$ seconds, on November 12th, 1906. The Voisin-Delagrange powered machine made brief hop-flights in 1907 (Fig. 53). In 1907 Henri Farman bought a Voisin machine and modified it over 1907 and 1908; he was (in November 1907) the first European to fly for more than a minute. In 1909—directly inspired by seeing the Wrights' lateral control in 1908—Farman introduced his own *Henry Farman III* (with ailerons), which became Europe's most popular biplane (Fig. 54). Meanwhile, in the United States, the Aerial Experiment Association in 1908 produced its *Red Wing*, the first potential domestic rival to the Wrights, which was derived from them (Fig. 55); then Glenn Curtiss, in 1909, evolved from this the first successful rivals to the Wrights, his *Golden Flier* and his Reims machine (Fig. 56).

47. Ferber's first, and unsuccessful, Wright-type glider: 1902.

48. Archdeacon's first, and unsuccessful, Wright-type glider: 1904.

49. Esnault-Pelterie's unsuccessful Wright-type glider, with elevons: 1904.

50. Ferber's Wright-type glider, with forward elevator and fixed tailplane: 1904.

51. The Voisin-Archdeacon float glider, combining Wright and Hargrave box-kite characteristics: 1905.

52. Santos-Dumont's powered biplane "14-*bis*" (modified), with ailerons added: 1906.

53. The Voisin-Delagrange I powered biplane: 1907.

54. The *Henry Farman III* powered biplane: 1909.

55. The Aerial Experiment Association's *Red Wing:* 1908.

56. The Curtiss Reims machine at Reims: August, 1909

Notes

THE WRIGHTS' PRIORITY IN POWERED FLYING

There has, over the years, grown up and become consolidated, what can only be called an "anti-Wright faction", which now perpetuates itself on both sides of the Atlantic. Its origins go far back in history, and derive from the disbelief and jealousy felt by the early French pioneers who, by the early years of this century, had become convinced that the homeland of the Montgolfiers possessed, so to say, an inalienable priority of invention—both in the past and in the future—of all things aeronautical. This attitude then spread to Britain and the United States.

The anti-Wright faction has basically two prongs of propaganda to its pitchfork; first, that a number of men succeeded in flying before the Wright brothers; and second, if such claims should by chance prove unfounded, then the Wrights, although they might have been the first to fly, achieved this priority in isolation, and never influenced anyone else; and, therefore, the aeroplane would have developed in the same way as it did, even if the Wrights had never existed. Neither of these theses is true, and the anti-Wright attitude concerning the Wrights' influence on Europe is dealt with in detail in my forthcoming Science Museum publication entitled *The Re-Birth of European Aviation.*

Here it is only necessary, I feel, to outline the claims that other pioneers flew before the Wrights.

The first serious claimant is the Frenchman Clément Ader. He certainly achieved, in 1890, an unassisted take-off with his steam-powered *Éole*, but the machine could not sustain itself, nor could it be controlled. His later claim to have flown for 300 metres in 1897, with his *Avion III*, is now known to have been mendacious.

Sir Hiram Maxim's giant biplane test-rig succeeded, in 1894, in lifting itself off its rails for a second or two before crashing; but no one pretends that it either flew properly, or was capable of flight.

The German-born American, Gustave Whitehead (Weisskopf), claimed he had made several powered flights in 1901 and 1902, in two monoplanes; the best of these were alleged to have been 1½ miles in 1901, and 7 miles—over Long Island Sound—in 1902. After a long and careful investigation, both the American authorities and ourselves have shown that such alleged flights were only flights of fancy.

It was claimed by his brother that the Scotsman Preston A. Watson made a powered flight in 1903; but after I had made a detailed analysis of this claim, the brother finally admitted he could no longer claim that Preston had made any powered flights in 1903, and claimed that he had only flown a glider.

Also in 1903, the German Karl Jatho made some powered hops in a semi-biplane, but they were neither sustained nor controlled, and came to nothing.

It is claimed that at some time before December of 1903, Richard Pearse made powered flights in New Zealand; but we now have two statements by Pearse himself saying that his first efforts to fly took place early in 1904, but were not successful, the machine, as he says being "uncontrollable".

THE WRIGHTS AND FRENCH ENGINES

In recent years a somewhat novel kind of denigration of the Wright Brothers has been put about on both sides of the Atlantic. As one writer has said, "The Wright aircraft was never able to leave the ground under its own power until the 1908 period, when it was fitted with a French engine, built in France by the firm of Bariquand and Marre of Paris".

For the record, it should be said that, (a) the outstanding flights made in the U.S.A. by Orville Wright in 1908 were made with a Dayton-built Wright engine; (b) Wilbur Wright flew brilliantly in France from August 8th to October 30th, also with a Dayton-built Wright engine; (c) after that date he used a standard Wright engine built under licence by the Paris firm named above. At no time did the Wrights use any engines other than those they designed themselves; in fact, they were not at all satisfied with the work done for them by Bariquand and Marre.

CLÉMENT ADER AND WING-WARPING

It was unfortunate that in 1903 the Wrights' friend Octave Chanute decided —in his writings about the Wright machines—to change the brothers' own term "wing twisting" to wing "warping": he never explained his reason for this inaccurate substitute, but the term "warping" soon became accepted everywhere. And when it came to be translated into French, it was rendered literally as "gauchir" (to warp) and "gauchissement" (warping), instead of using the proper translation of the Wrights' word to "twist", which was "tordre". Thus it came about that as soon as Clément Ader saw the two words for warping in French he immediately claimed he had invented wing-warping, and pointed to his aircraft patent of 1890 where both words had been used. But when I examined his 1890 patent and the explicit illustrations to the text, it became all too clear that what Ader had designed his wings to do was not to change their angles of incidence—which the Wrights did—but simply to flex the tips up and down, with no change of incidence. Although Ader has been revealed in a very bad light as to his flight-claims, it might just be argued in mitigation, in this question of warping, that he did not really understand what the word "warp" meant when applied to the Wrights' wings.

BIRTHPLACES AND DATES

Wilbur Wright was born near Millville (Indiana) on April 16, 1867: he died of typhoid fever on May 30, 1912 at Dayton (Ohio).

Orville Wright was born in Dayton (Ohio) on August 19, 1871: he died on January 30, 1948 at Dayton.

EYEWITNESS ACCOUNT OF A WRIGHT FLIGHT (1904)

It is often said that we are asked to accept the Wright brothers' first powered flights in 1903, 1904 and 1905 solely on their own testimony: this was also the viewpoint adopted by most of the early French pioneers, who refused to believe the unpalatable news of these flights which was reported from America. Apart from the fact that the Wrights have never once been found guilty of knowingly telling an untruth, and despite the various local witnesses they named for those three years, it is not generally known that there exists an impeccable eyewitness of their flying in 1904—thus long before any serious European claim to having flown—a witness who was so deeply impressed by what he saw, that he not only committed the whole experience to paper, but published it. This person was Mr. Amos I. Root, proprietor and editor of the respected apiarists' journal called *Gleanings in Bee Culture*, which still flourishes today under the proprietorship of Root's descendant, and is still being published in Medina, Ohio, where it started. Amos Root was lucky enough to witness the Wrights' first circle on September 20th, 1904, the world's first circle by an aeroplane. His long and detailed description of the Wrights and their flying was published in the issue of the *Gleanings* dated January 1st, 1905, and occupied four double-column pages of the magazine. It has the added distinction of being the first eyewitness account of a powered flight in the history of the world. Here are some brief extracts:

"God in His great mercy has permitted me to be, at least somewhat, instumental in ushering in and introducing to the great wide world an invention that may outrank electric cars, the automobiles, and all other methods of travel, and one which may fairly take a place beside the telephone and wireless telegraphy. . . . It was my privilege, on the 20th day of September, 1904, to see the first successful trip of an airship, without a balloon to sustain it, that the world has ever made, that is, to turn the corners and come back to the starting-point. . . . When it first turned that circle, and came near the starting-point, I was right in front of it; and I said then, and I believe still, it was one of the grandest sights, if not the grandest sight, of my life. Imagine a locomotive that has left its track and is climbing up in the air right towards you—a locomotive without any wheels, we will say, but with white wings instead. Well, now, imagine this white locomotive, with wings that spread 20 feet each way, coming right towards you with a tremendous flap of its propellers, and you will have something like what I saw. The younger brother bade me move to one side for fear it might come down suddenly; but I tell you, friends, the sensation that one feels in such a crisis is something hard to describe. . . . When Columbus discovered America he did not know what the outcome would be, and no one at that time knew. In a like manner these two brothers have probably not even a faint glimpse of what their discovery is going to bring to the children of men. No one living can give a guess of what is coming along this line. . . . Possibly we may be able to fly *over* the north pole, even if we should *not* succeed in tacking the 'stars and stripes' to its uppermost end."

FURTHER READING

The literature on the Wrights is so voluminous, and is spread through so many books and periodicals, that it is suggested that the reader who comes new to the subject might progress in four phases, as follows:

1. For a general survey of the Wrights' work; descriptions of their early aircraft; their influence on European aviation; etc.

C. H. GIBBS-SMITH. **Aviation: an historical Survey.**
(Science Museum publication.) London: H.M. Stationery Office, 1970.

2. For a general biography of the brothers, authorised by Orville Wright, but not very reliable on technical questions:

FRED C. KELLY. **The Wright Brothers.** New York, 1943. (London, 1944.)

3. For investigation into every aspect of the personal and technical lives of the Wrights, there is a monumental work available in two volumes and covering over 1,300 pages, with over 250 illustrations:

The Papers of Wilbur and Orville Wright.
Edited by MARVIN W. MCFARLAND. 2 vols. New York, 1953.

4. For special investigation into particular aspects and problems, there is a 23-page bibliography at the end of volume 2 of the above work, which will direct the reader to every important source and commentary. Added to this there has now been published an exhaustive Wright bibliography, as follows:

WASHINGTON: LIBRARY OF CONGRESS. **Wilbur and Orville Wright: a Bibliography (etc.)** Compiled by A. G. Renstrom. Washington, 1968.

Grateful acknowledgement is made to Marvin W. McFarland, Chief of Science and Technology, Library of Congress, Washington, D.C.; and to the authorities of the National Air and Space Museum (Smithsonian Institution), Washington, D.C.

Printed in England for Her Majesty's Stationery Office
by Eyre and Spottiswoode at Grosvenor Press Portsmouth

Dd. 500644 K96 9/72